Find a wide bowl.

Find a tall bottle.

Find a cup.

Which holds more?

Find 2 more pots.

cupfuls

Which holds less?

- Can use language of comparison (holds more/less than).

Notes/date:

P.A

- Can understand position words.

Notes/date:

mouse

rabbit hutch

cat

mouse cage

rabbit

fish

cat

cat basket

cat

Who has no home?

- Complete simple mapping. Notes/date:

2.3

8						
7						
6						
5						
4						
3						
2						
1						
	cat	dog	fish	mouse	rabbit	

- Can create a simple block graph.   Notes/date:
- Can interpret block graph.

2.3

Make 4 of each.

Make 5 currants on each bun.

- Can count objects to 5. Notes/date:
- Can check total.

2.4

Draw the coins.

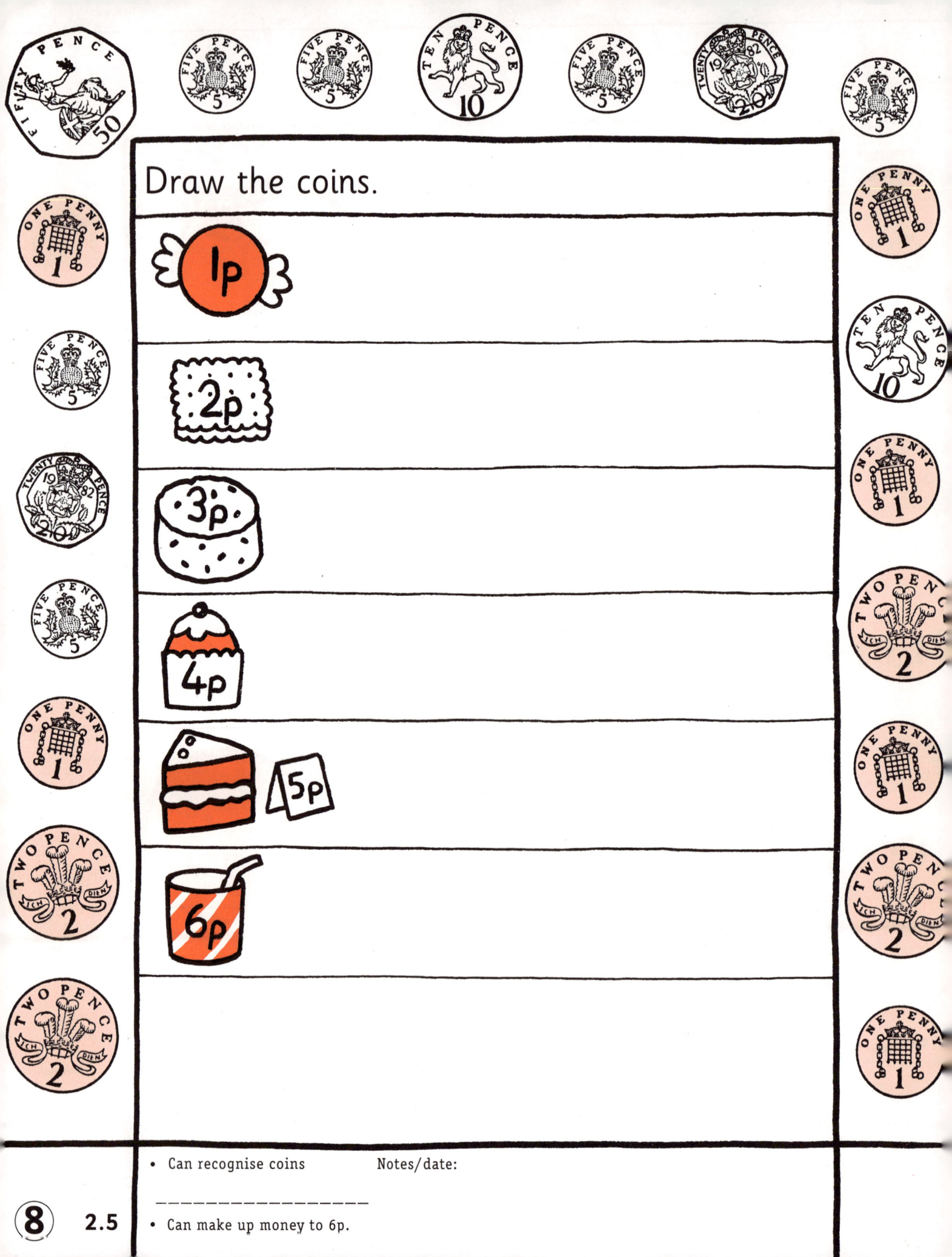

1p

2p

3p

4p

5p

6p

- Can recognise coins   Notes/date:
  _____
- Can make up money to 6p.

2.5

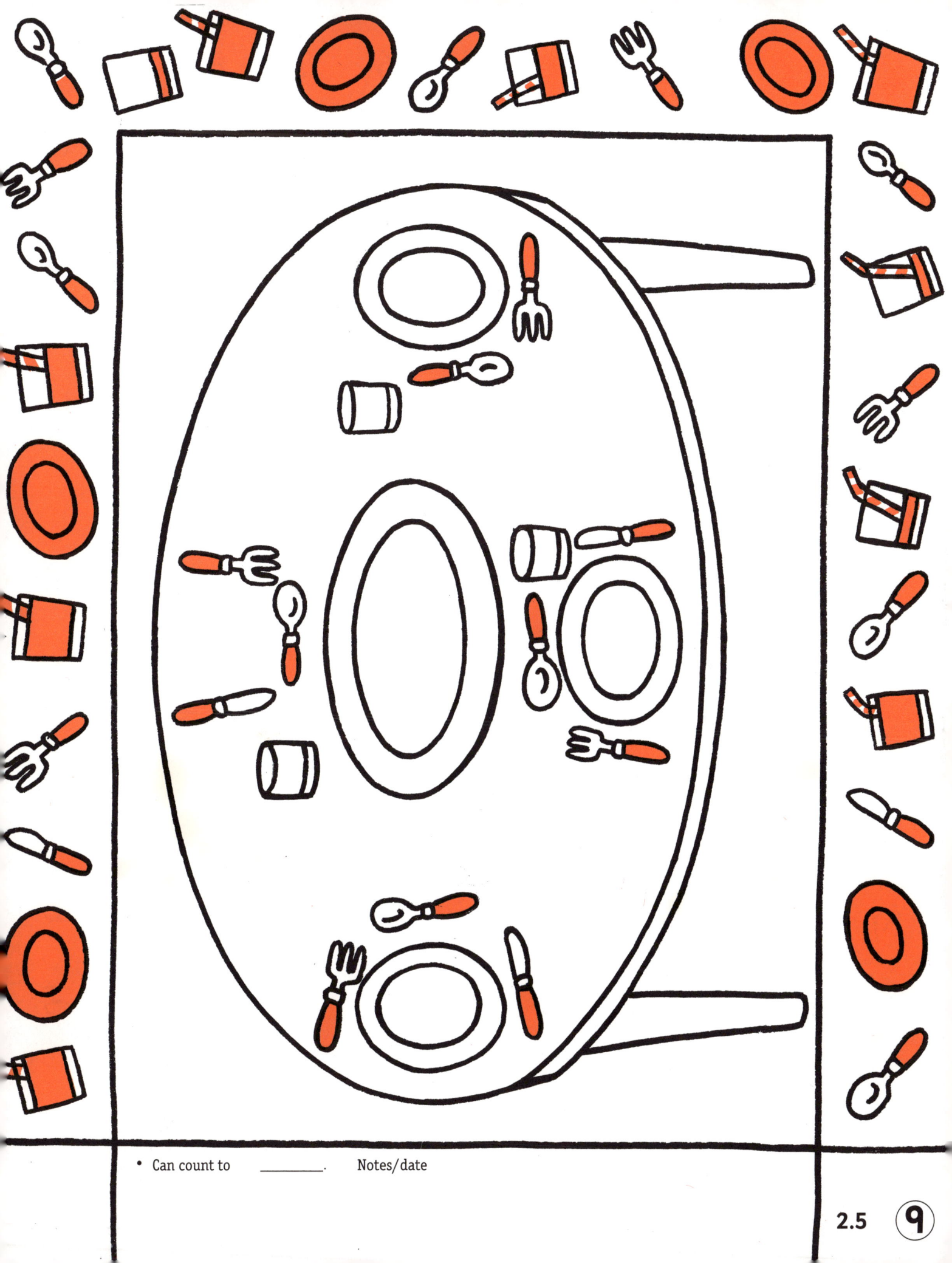

- Can count to _____.   Notes/date

2.5

5

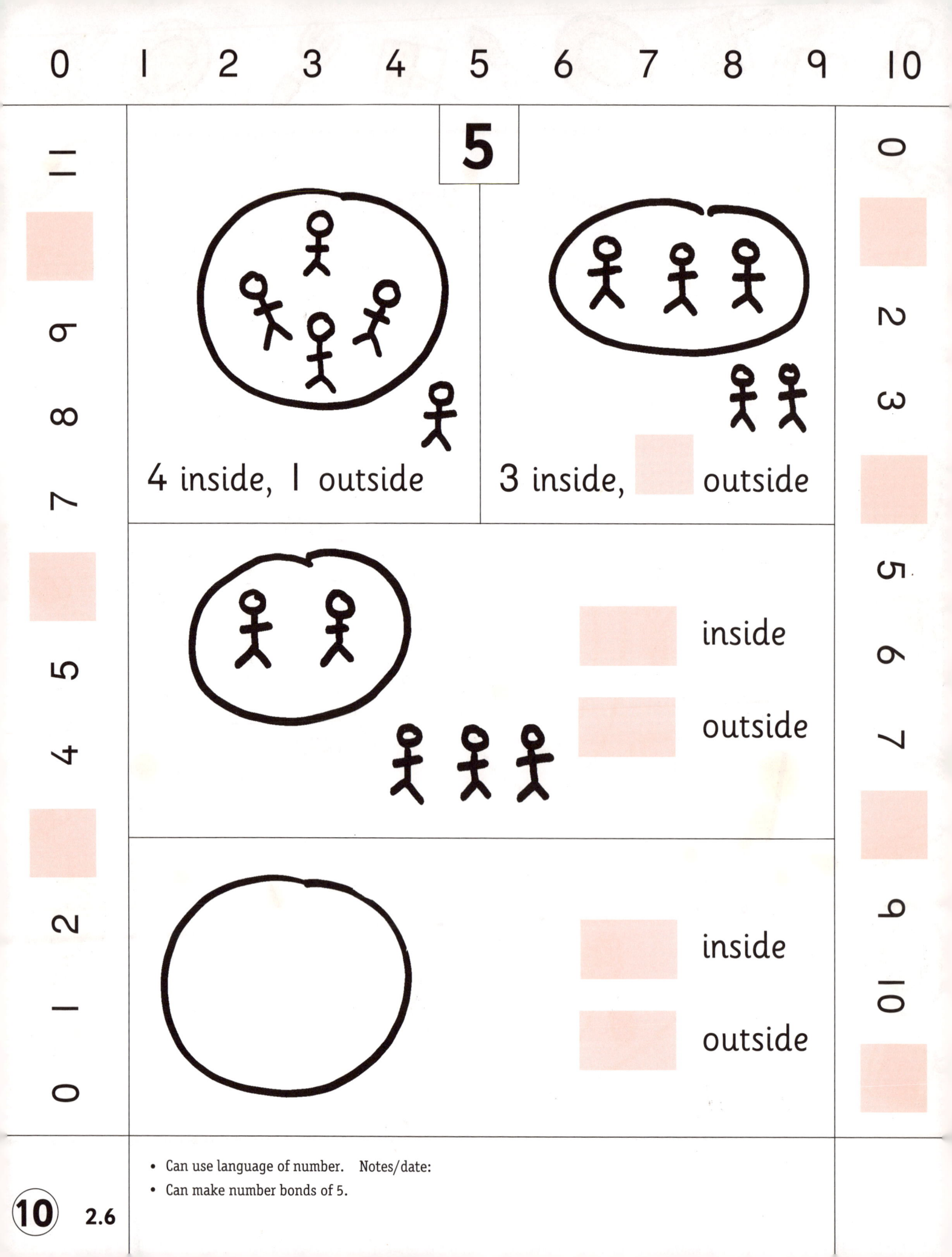

4 inside, 1 outside

3 inside, ☐ outside

☐ inside
☐ outside

☐ inside
☐ outside

- Can use language of number.   Notes/date:
- Can make number bonds of 5.

Write	draw
1 1 one	
2 2 two	
3 3 three	
4 4 four	
5 5 five	
6 6 six	
7 7 seven	
8 8 eight	
9 9 nine	
10 10 ten	

• Can count to 10   Notes/date

P.A9

Draw what you found out.

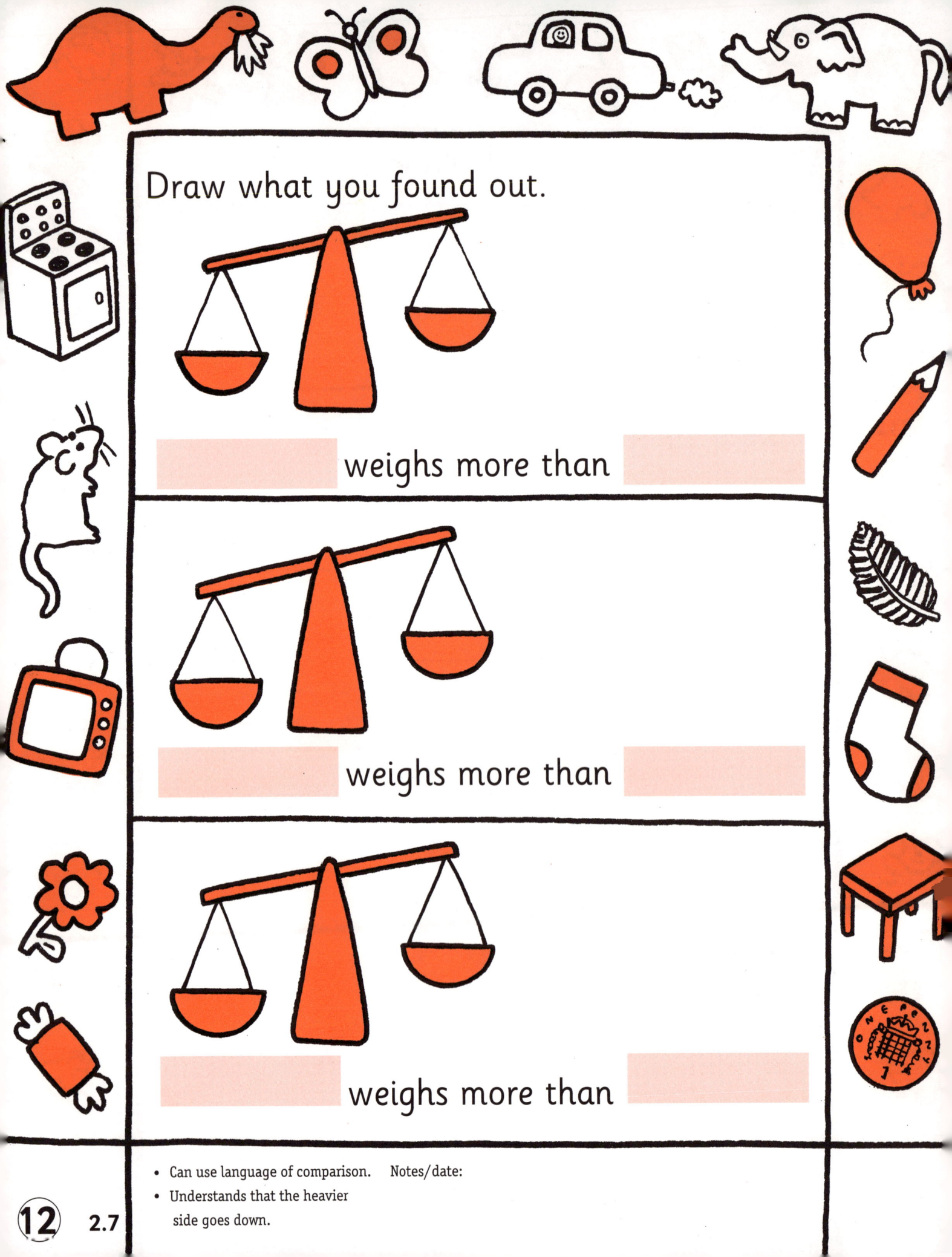

_____ weighs more than _____

_____ weighs more than _____

_____ weighs more than _____

- Can use language of comparison.   Notes/date:
- Understands that the heavier side goes down.

You need   pebbles   skipping rope

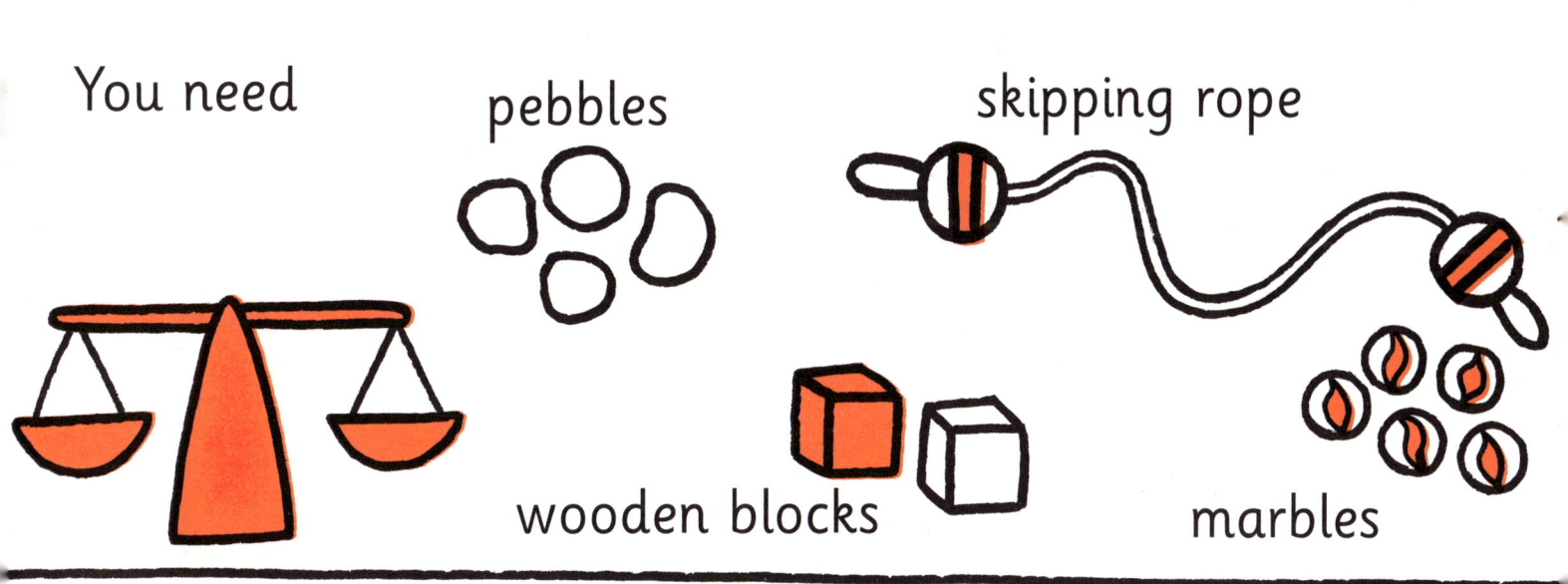

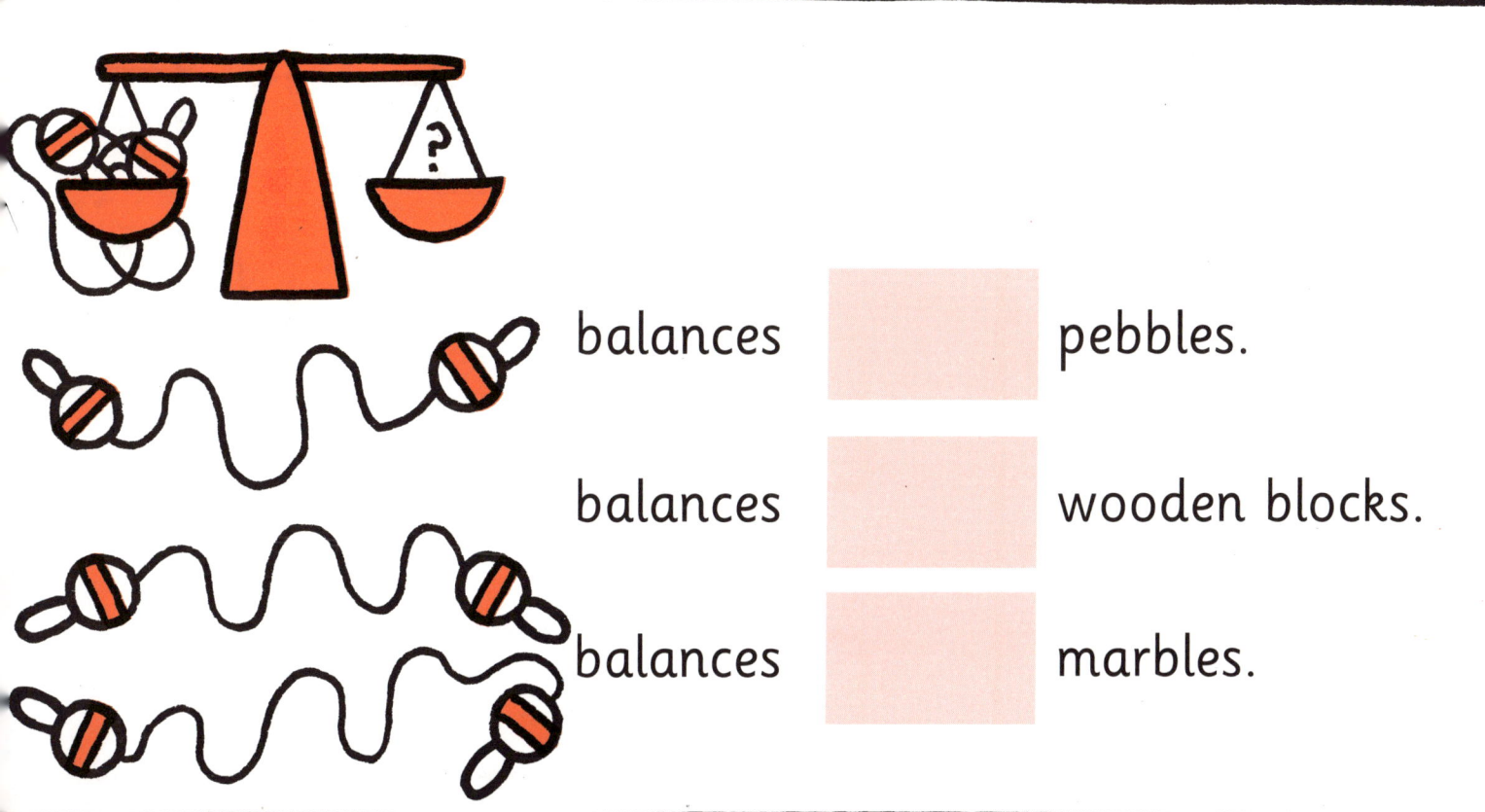

wooden blocks   marbles

balances ▢ pebbles.

balances ▢ wooden blocks.

balances ▢ marbles.

Do some more.

........................balances........................

........................balances........................

- Can balance the scales.   Notes/date:
- Can talk about work.

Little Ted goes to the picnic.

Draw a path in red.

Draw a longer path in blue.

- Can describe a route.   Notes/date:
- Understands longer.

Now I can

date

I played

maths games.
date

I can count to
date

My best pattern was

date

PUBLISHED BY THE PRESS SYNDICATE OF THE UNIVERSITY OF CAMBRIDGE
The Pitt Building, Trumpington Street, Cambridge CB2 1RP, United Kingdom

CAMBRIDGE UNIVERSITY PRESS
The Edinburgh Building, Cambridge CB2 2RU, United Kingdom
40 West 20th Street, New York, NY 10011-4211, USA
10 Stamford Road, Oakleigh, Melbourne 3166, Australia

Sue Atkinson   Sharon Harrison
Lynne McClure   Donna Williams

Illustrated by Cathy Baxter

© Cambridge University Press 1995

First published 1995
Third printing 1997

Printed in the United Kingdom at the University Press, Cambridge

Key

1 – ◯

2 – nothing

3 – △

4 – ☐ or ▭

5 – ⬠

6 – ⬡

CAMBRIDGE UNIVERSITY PRESS

ISBN 0-521-47582-1

9 780521 475822